Conversations on Optimal Transport

Luigi Ambrosio • Alfio Quarteroni
Editors

Conversations on Optimal Transport

 Springer

Editors
Luigi Ambrosio
Scuola Normale Superiore
Pisa, Italy

Alfio Quarteroni
Politecnico di Milano
Milano, Italy

EPFL, Lausanne, Switzerland

ISBN 978-3-031-51684-9   ISBN 978-3-031-51685-6   (eBook)
https://doi.org/10.1007/978-3-031-51685-6

Specifically created for this book via an AI tool

This Springer imprint is published by the registered company Springer Nature Switzerland AG
The registered company address is: Gewerbestrasse 11, 6330 Cham, Switzerland

If disposing of this product, please recycle the paper.

Foreword

From Webinars to Podcasts to Books

This book gems from an idea born in 2020 during the lockdown that affected the whole world, changing our lives, our habits, our daily routines, and our business models for ever.

At that time, the reader might remember that all the meetings and conferences planned had to be cancelled or postponed. Therefore, with some colleagues from the mathematics journal department at Springer, we decided to start a series of SN conferences, which basically consisted of offering to the journals and books Editorial Boards we cooperate with the possibility of hosting and coordinating virtual conferences, so as not to interrupt the dissemination of research results.

In the following years, when everything returned to the so-called "new normality", we stopped organizing online conferences, and we further developed the concept by launching the **SN webinars**, a series of branded virtual interviews and webinars involving either top Springer authors or Key Opinion Leaders in a certain field.

In this contest, I (as Executive Editor, Mathematics, Books division at Springer) involved the Board members of a successful Springer textbook series in mathematics—UNITEXT—which is rather well-known at an international level. We decided to start a series of interviews that are handled by the UNITEXT Board members, who would rotate to interview top experts in their field.

The starting point is the topic faced in a textbook, and the discussion then expands and covers more general topics related to the research and the personal history of the guest speaker. The audience is invited to interact with the speakers, and it is expected to ask questions and further develop the discussion.

The result is a series of video recordings, mainly focusing on a very hot topic: Optimal Transport. These conversations involve not only authors from the UNITEXT series, but also members of the book series' Editorial Board. Additionally, they feature prominent figures in the field, including a Field Medalist.

At that point, we realized that we had high-quality material in our hands; therefore, we had the idea to generate new content working on these video recordings and creating audio **PODCASTS** out of them. These PODCASTS are already available on our dedicated channel: https://springer-mathpodcast.buzzsprout.com/.

The following step bloomed rather naturally: we moved forward to publish a book out of the PODCASTs. Subsequently, advanced AI tools were employed—under human supervision—to transcribe the audios and edit them for better readability.

It is important to note that the content in each format—the interviews, the PODCASTS, and the book—is self-contained and not a mere adaptation from one medium to another. Instead, it represents an independent exploration of the subject matter.

This book can be considered the result of excellent synergies between Springer Editors and prominent Springer authors, with the support of sophisticated artificial intelligence (AI) tools.

For a few months, we have all been talking or hearing about the AI ability to answer complex questions, write songs, generate codes, translate languages, and—as in this specific case—produce scripts from audio recordings.

Because of the infinite multiple uses that open before us, many perceive the impact of AI as inevitable and potentially threatening. Nothing new: every time in human history when a new technology has entered our lives, it has led to a revolution: the agricultural, industrial, and information technology ones. These revolutions resulted in the elimination of heavy and repetitive work with little cognitive content, but at the same time they created new jobs. Now, thanks to what is called *machine learning*—which focuses on the use of data and algorithms to imitate the way that humans learn and steadily enhances its precision over time—we can automate tasks and processes, reducing the time and effort needed for various activities. Human intelligence plays a unique role here; artificial intelligence algorithms can solve problems, but their precise and accurate formulation is vital and is in our hands. Therefore, the human touch is fundamental in this process, concentrating on problem formulation to obtain an effective result.

Let us contextualize this concept: we managed to create the image on the cover using one of the many AI tools available, DALL-E 3. We asked it to represent some Italian mathematicians talking about some specific topics related to mathematics, AI, and Optimal Transport. We refined and rephrased the request 20 times to obtain a result close to what we had in mind. Here is the request we originally made, which brought to the final option which you can find printed on the book cover:

"Artistic flat design where a study room and an Italian landscape merge without a hard separation, connected by a wavy 'optimal transport' curve. The study is detailed with vintage elements and AI tools. The countryside captures the essence of Italy with mathematicians in contemplation. AI digital elements and mathematical icons transition smoothly between the two scenes."

We applied the same method when we asked the tool to revise the text from a linguistic perspective, deleting any Italian accent, removing any mistake and any distortion of the original meaning: every new request produced different results, which were carefully checked, revised, and then approved. It is of the outmost importance to emphasize that a human copy-editor was involved, and the authors revised the final copy-edited texts before publication.

We hope the reader will appreciate this booklet and the concept behind it, which allowed us to test some AI tools, leaving us the role of creating and supervising the whole process; and, ultimately, of having fun as well.

***

I would like to express my gratitude to Alfio and Luigi, with whom I have collaborated for several years in the exploration of new books and authors and in the publication of high-quality content. Despite their myriad and more pressing daily responsibilities, they have embraced to be involved in this new project.

Special thanks are due to my esteemed and dearest colleagues Marc Strauss, Jan Holland (and to Clemens Heine, who sadly passed away last year) of the Springer Maths Journal team. Since 2020, we have developed the **SN**

webinars project, and are still cooperating to further develop the concept in mutual synergy.

I express my gratitude to Robinson Dos Santos of the Springer Maths Books team, who is also responsible for the Springer Maths PODCAST channel, for having enthusiastically accepted to contribute to this idea by transforming original videos into captivating audio recordings.

Last, but not certainly least, my gratitude goes to Jonas Spies of the Books Publishing Solutions division. As an exceptionally skilled colleague engaged in developing AI-related projects and testing cutting-edge tools, Jonas played a crucial role in concretely creating this booklet.

To each of you, my sincere thanks: you contributed to make this dream come true!

Milan, Italy   Francesca Bonadei

Preface

This work is closely tied to the renowned mathematics text-book series known as UNITEXT. UNITEXT is a series of textbooks tailored for university students pursuing bachelor's or master's degrees.

What sets this particular book apart in the Springer collection is its unique origin. It has been crafted through a meticulous process involving interviews with world-class mathematicians in leading institutions: the Eidgenössische Technische Hochschule in Zürich (ETH), the Scuola Normale Superiore in Pisa (SNS), the Scuola Internazionale Superiore di Studi Avanzati (SISSA) in Trieste. These interviews were originally broadcasted as online events on the Cassyni platform. Afterwards, the videos were transcribed into audios and then converted into text using sophisticated AI tools.

The interviews featured in this book revolve around a highly relevant and engaging topic: Optimal Transport.

Optimal Transport is a research field originated from a problem presented in a famous Memoire presented by Gaspard Monge in 1781. Its development, by the lack of proper mathematical tools, come only much later, around 1940, with the birth of the modern theory of Probability and the pioneering work of Leonid Kantorovich. In the last few decades, thanks to independent and almost simultaneous contributions, deep connections of this theory with many more research fields emerged: Elliptic Partial Differential Equations, Fluid Dynamics, Geometric and Functional Inequalities, Metric Geometry. Even more recently, the metric generated by Optimal Transportation is used in the analysis of images and as a discriminator in artificial intelligence algorithms. The interviewees have played an important role in many of these developments.

What truly sets this book apart is the exceptional quality of both the interviewers and interviewees. It provides readers with a snapshot of remarkable vitality and freshness, guaranteed to captivate and engage anyone with an interest in contemporary mathematical issues, and in future developments in the field.

Pisa, Italy   Luigi Ambrosio
Milano, Italy   Alfio Quarteroni
Lausanne, Switzerland
November 2023

Contents

Editors and Contributors

About the Editors

Luigi Ambrosio is a Professor of Mathematical Analysis, a former student of the Scuola Normale Superiore and presently its Director. His research interests include calculus of variations, geometric measure theory, optimal transport and analysis in metric spaces. For his scientific achievements, he has been awarded several prizes, in particular the Fermat prize in 2003 and the Balzan Prize in 2019.

Alfio Quarteroni is Professor Emeritus at Politecnico of Milan (Italy) and at EPFL (Swiss Federal Institute of Technology), Lausanne (Switzerland). He is the founder (and first director) of MOX at Politecnico of Milan (2002) and MATHICSE at EPFL, Lausanne (2010). He is co-founder (and President) of MOXOFF, a spin-off company at Politecnico of Milan (2010). He has been an invited or plenary speaker in more than 300 International Conferences and Academic Departments. For his scientific achieve-

ments, he has been awarded several prizes; the last one is
ICIAM Lagrange Prize 2023.

Contributors

Luigi Ambrosio Scuola Normale Superiore, Pisa, Italy

Camillo De Lellis School of Mathematics, Institute for Advanced
Study, Princeton, NJ, USA

Alessio Figalli Department of Mathematics, ETH Zurich, Zurich,
Switzerland

Nicola Gigli SISSA, Trieste, Italy

Alfio Quarteroni Politecnico di Milano, Milan, Italy
EPFL, Lausanne, Switzerland

Talking about Optimal Transport

Luigi Ambrosio and Alfio Quarteroni

1   Introduction

In this session, we have the pleasure of hosting Luigi
Ambrosio, a professor at the Scuola Normale Superiore in
Pisa, Italy, as our guest. Professor Ambrosio, who recently
co-authored the new textbook, **Lectures on Optimal
Transport**, with Elia Bruè and Daniele Semola, engages in

The original version of this chapter has been revised. A correction to this chapter
can be found at https://doi.org/10.1007/978-3-031-51685-6_4

L. Ambrosio (✉)
Scuola Normale Superiore, Pisa, Italy
e-mail: luigi.ambrosio@sns.it

A. Quarteroni
Politecnico di Milano, Milan, Italy
EPFL, Lausanne, Switzerland
e-mail: alfio.quarteroni@polimi.it

a lively conversation with Alfio Quarteroni, a professor at Politecnico di Milano.

During this special interview, which is hosted by Professor Quarteroni, Professor Ambrosio shares his insights on topics such as optimal transport, the evolution of mathematics in Italy, strategies for motivating young mathematicians, and unsolved problems in his field.

Originally aired by the Springer Nature webinar series, this interview has been specifically adapted for the podcast format.

Alfio Quarteroni:
Good morning, good afternoon, and good evening to everybody. It's a pleasure to interview such a great colleague and author. Let me begin by discussing your book published by Springer, titled "Lectures on Optimal Transport." Can you tell us what it is about and what made you decide to write it? Also, according to your own perspective, what makes it special and different from other books on the same topic?

Luigi Ambrosio:
Thank you, Alfio. First of all, let me say that I'm delighted to be interviewed by you about mathematics. I became fascinated by the topic of optimal transport approximately 20–23 years ago when the subject was experiencing rapid growth. To recount the story, I remember being invited to a summer school in Madeira organized by some colleagues in Pavia. At that time, I was still in Pavia or had left just a few years prior. I was initially supposed to speak about other topics in the calculus of variations, but I was captivated by this new subject and immediately decided to change the focus of my lectures to optimal transport. Consequently, I quickly began writing a short set of notes, which we still use to this day.

Starting from that experience, I regularly offered courses on that topic at the Scuola, perhaps every 3 to 4 years. In the meantime, the subject continued to grow, and my research in this field expanded into different areas. As we will see, this began with the existence of optimal maps, but then moved to PDEs, geometric analysis, and so on. Gradually, I improved this set of lectures until about 4 years ago, when I asked the students in my course to help me write the first draft. Finally, 2 years ago, with Elia Bruè and Daniele Semola, we decided to give the book its final form. In essence, this book is the culmination of many years of teaching experience on the subject.

Alfio Quarteroni:
I can only imagine the excitement Elia and Daniele must have felt when you proposed writing this book together. You titled this book "Lectures on," and I believe we now understand why you chose this specific title. It reflects a stratification of knowledge and the gradual development of your teaching experience on this subject. The book, which, as usual, is presented in a pedagogical manner, can be used by beginners, allowing them to dive in and explore the subject with the depth and clarity for which you are widely recognized.

In the preface, you mention that after a comprehensive introduction to the classical theory, which includes the formulation of the problem, duality, necessary and sufficient optimality conditions, and the existence of optimal maps in Euclidean spaces as well as Riemannian manifolds, you then focus on applications to geometric and functional inequalities and to PDEs. Could you provide some examples that might help our audience better understand the topic?

Luigi Ambrosio:

Yes, perhaps I should first address the title. Sometimes, when I want to teach a course on a particular subject, I think, "Well, I'll use this book as a guide for the course." However, if it is not a textbook but rather a research book, this can become nearly impossible. One might find oneself jumping from one section to another or consulting multiple books simultaneously, and then...

Alfio Quarteroni:

And students tend to be somewhat unhappy with that approach, right?

Luigi Ambrosio:

In some sense, I organized this book in a way that, at least ideally, each chapter represents a single lecture. As a result, one can genuinely create a course using this book. Of course, it would be a fairly advanced course, suitable for PhD students or capable undergraduate students. Ideally, you could design a 40- or 50-hour course, devoting 2 h to each chapter. This allows you to use the book for a course without needing to jump around too much.

As I mentioned in the introduction, this subject has been growing at a rapid and somewhat unpredictable rate. For example, when I wrote my book with Gigli and Savarè on optimal transport and its applications to PDEs around 2005, it was not entirely clear that there would be further applications of this theory. The applications to PDEs and functional inequalities were apparent, but the applications to geometric analysis that have been the main focus of the last 10–12 years were completely unforeseen. When this growth began, I recall the immense and commendable effort by Cédric Villani to write his comprehensive monograph, titled "Optimal Transport, Old and New," which is also a Springer publication. I believe his book was the last attempt to write something comprehensive about the theory, akin to an encyclopedia.

However, even after that, in the last 12–13 years, the subject has developed significantly. It is now almost impossible to cover all aspects of optimal transport in a single book. So, what is the best strategy? As I wrote in my book, you receive the basics, hints, and some main directions, but if you are interested in specific applications, you should consult either the literature or other more specialized textbooks. For instance, there are lecture notes written by Alessio Figalli and a book titled "Optimal Transport for Applied Mathematicians" by Filippo Santambrogio, another former student of the Scuola Normale. There are now many branches of the subject, but my idea is that with this course, based on the feedback from students, you can genuinely spark their interest in the topic.

Alfio Quarteroni:
As you mentioned earlier, this subject has been around for roughly 25 years, right? You told us about the origin of your interest in the subject, dating back to the Madeira Conference. It's an old problem, associated with famous names like Monge, Kantorovich, and many others. You also mentioned progress in recent years. Can you speculate on the reasons behind the booming popularity and importance of this subject in recent times?

Luigi Ambrosio:
First, I should explain why, from my perspective, this subject remained stagnant for a long time. The first version of the optimal transport problem dates back to Monge during the years of the French Revolution. Although there were heuristics about the behavior of solutions, almost no rigor-

ous results emerged until Kantorovich's work in the 1940s – a jump of nearly two centuries. It's an interesting story, which can also be found in a paper A.M.Vershik in the Mathematical Intelligencer. Apparently, when Kantorovich proposed his version of the optimal transport problem related to dynamic programming and economics (for which he was awarded the Nobel Prize in Economics), he was not aware of Monge's original version and only discovered it later. This serves as a fascinating example of how mathematical ideas can sometimes disappear and re-emerge like underground rivers.

By Kantorovich's time, the stage was set for the first developments, as the tools of probability theory were already available, unlike in Monge's era. One could say that the optimal transport problem is a very classical problem of the calculus of variations. When compared to the more well-known problems of the calculus of variations of Lagrange, Euler, and others, including the contributions from the Italian school such as Tonelli, Lebesgue, and Cesari, one might wonder why the optimal transport problem did not receive as much attention. I believe the main reason was the absence of mathematical tools until the advent of probability theory.

As for what happened in more recent years, there are two or three founding fathers of these new developments. It began with the work of Robert J. McCann.

McCann was the first to introduce a new way of interpolating between probability densities using optimal transport. In a more geometric setting, it was also Felix Otto who understood how certain PDEs, particularly the porous medium equations, could be approached using optimal transportation in more geometric terms. The third founding father of the most recent developments is Yann Brenier, who established strong connections between optimal transport and the theory of fluid mechanics. These developments began around 1990.

Perhaps the simplest explanation I can provide without using formulas is that functional analysis has been very successful in tackling many PDEs in fluid mechanics, such as the heat equation, the Fokker–Planck equation, the transport equation, and other variants of PDEs in mathematical physics. However, we sometimes overlook the elementary fact that, for many PDEs in mathematical physics, the function ρ represents a probability density, which can be viewed as a cloud of points. It is more natural, from a fundamental perspective, to connect the probability densities by considering them as clouds of points rather than functions. This is the essence of optimal transportation.

Of course, functional analysis remains crucial, but with optimal transportation, we rediscover the Lagrangian interpretation to some extent. Many of the most significant developments in geometric analysis result from identifying concepts that can be stated in Lagrangian terms with those that can be defined in Eulerian terms. For example, in calculus, one can compute the modulus of the gradient of a function as the limit of the modulus of difference quotients along all directions – let's call this an Eulerian perspective, with a slight abuse. Alternatively, one can compute the modulus of the gradient of a function by examining the slope of these functions along all Lipschitz curves to normalize the velocity – this is a Lagrangian perspective. It is elementary to show in Euclidean space or on smooth manifolds that these two concepts are the same. However, their coincidence is a more universal phenomenon, which leads to developments in geometric analysis for spaces that are not smooth, specifically in the category of metric measure spaces.

Alfio Quarteroni:

That's a fascinating perspective, indeed. There are a couple of comments that come to mind. First, substantial progress in mathematics often involves reinterpreting concepts from different angles using various notations and tools. Second, you mentioned the three pillars in the evolution of the concept: probability with McCann, geometry/PDEs with Otto, and fluid dynamics with Brenier. This highlights the unity of mathematics and the lack of distinction between pure and applied, as well as between fundamental and specialized knowledge. The more one knows in terms of mathematics, the better equipped one is to allow theory to evolve and find suitable environments for new applications, as is the case with optimal transport. Thank you very much for this insightful discussion.

On a more personal note, you mentioned Alessio Figalli, who was awarded the Fields Medal in 2018 for his contribution to optimal transport. Alessio is just one of several outstanding students who obtained their PhD under your guidance; you also mentioned Filippo Santambrogio earlier. Can you explain the secret behind your success in attracting brilliant students and helping them become outstanding scientists? What is your recipe?

Luigi Ambrosio:

Answering that question is difficult. First of all, I would say that there is an element of luck involved. In a sense, I feel as though I am in the middle of a wonderful story, with Ennio De Giorgi before me and students like Alessio Figalli, Camillo De Lellis, Guido De Philippis, Nicola Gigli, and many others following my lead. So, what is the secret? I'm not entirely sure. However, I believe the special environment in Pisa, with its long-standing tradition in mathematics and revitalization by people like De Giorgi, Andreotti, and Stampacchia, plays a role. In some sense, I am a part of this long tradition, and the Scuola Normale is an essential component of it.

At the Scuola Normale, we select our students in mathematics and other disciplines on a national basis, and I would like to emphasize that our competition is open to students worldwide. The small number of students creates a cohesive community where they work together and have close interactions with professors. Starting with already talented students, we expose them to in-depth courses during their first 2 years. The idea is not to provide them with an extensive array of knowledge immediately, but rather to help them delve deep into the subjects. For instance, they learn the foundations of calculus and axiomatic set theory, and they engage in numerous exercises. This approach, which involves exposing students to elementary yet challenging exercises, is perhaps the main difference between the Scuola Normale and other universities. This strong training typically takes place during the first 2 years of their education.

When considering the students at the Scuola Normale, it's important to note that even though they are all very talented, their paths aren't identical. Some prefer a slower trajectory, while others, like Camillo and Alessio, benefit from early exposure to research and meetings. It's crucial to find the right balance and avoid putting too much pressure on students, as this could be detrimental. In summary, the combination of a strong preliminary selection, a nurturing environment, and early exposure to advanced research when appropriate contributes to the success of these students.

Alfio Quarteroni:

That's an interesting perspective, as nowadays, younger generations are used to exploring various subjects. You're suggesting that the key at the Scuola Normale is to encourage students to delve deeper and use elementary tools to achieve profound explanations and results. It's a different approach that may seem old-fashioned, but it's likely part of the success.

Luigi Ambrosio:

In connection with what you're saying, we do face pressure from various directions to be more interdisciplinary and open to different perspectives. However, we usually maintain a conservative approach during the first 2 years.

Alfio Quarteroni:

Okay.

Luigi Ambrosio:

In the third year, we incorporate the ideas you mentioned.

Alfio Quarteroni:

Then you open up students' minds and encourage them to explore.

Luigi Ambrosio:

Yes, that happens much more now than it did, for example, 20 years ago.

Alfio Quarteroni:

I understand that the aim is to establish a solid foundation in mathematics, allowing students to be more open and receptive to external influences, find connections, and explore various fields. However, I'd like to touch on another point. I notice the Italian flag and the EU plaque behind you. Until about 15 or 20 years ago, it was relatively common for Scuola Normale graduates to become professors at Italian universities. Nowadays, we see that most of them leave Italy and often become brilliant scientists in other countries. Can you explain the reasons behind this diaspora? While some degree of international movement is expected, when the percentages are too high, it may seem unreasonable. Do you have any recommendations or advice to mitigate this process, which, to some extent, is entirely normal?

Luigi Ambrosio:
First, let me say that I am deeply committed to addressing this issue, even discussing it with the Ministry. When I became the director of the Scuola Normale, it was one of the first topics I raised. Particularly for the Scuola Normale, this situation is a waste. We estimated a few years ago – though this figure may not be realistic anymore – that training a student at the Scuola Normale costs around €100,000 or even more. It's like building an excellent car and then giving it away for free. There is a clear contradiction.

The reasons behind this issue are more or less apparent: there is a salary gap and basic uncertainty about career prospects due to the lack of stable rules in Italy. In contrast, the French system has regular yearly admissions of positions and well-established rules that may change slightly but not dramatically. In Italy, funding for research, which is also related to the training of young scientists, is rather random. These are the fundamental reasons.

However, I must say that in recent years, while the problem hasn't been solved completely, I have seen some effective actions taken by the last two ministers of Research and University, Manfredi and the current minister, Messa. Now, we have the possibility to hire scientists at earlier stages of their careers through simplified procedures if they have obtained an ERC grant or even a seal of excellence. This recognition in Italy allows them to potentially have a faster career progression. For example, Elia Bruè, one of the authors of this book, obtained his PhD 3 years ago, spent a year and a half at the Institute for Advanced Study in Princeton, and recently secured an RTD-B position at Bocconi University. I'm delighted to see that Elia returned to Italy, and I'm confident that if Daniele, who has been in Oxford and is now in Zurich, decides to come back, he will find opportunities using these tools, such as the Levi–Montalcini grant.

Of course, the numbers are still small compared to the global phenomenon you mentioned. We should not only be concerned about the top talents leaving the country; the problem is that in the current situation, many other researchers leave as well, which is even worse and increases the overall numbers. Nonetheless, I have noticed some positive recent developments in this regard.

Alfio Quarteroni:
I'm glad to see that you're becoming optimistic about this issue, which is so important for our country. As one of the top non-linear analysts worldwide, you have opened new mathematical avenues in the field of variational analysis and geometric partial differential equations. Can you give us some ideas on which, in your opinion, are the most challenging open problems in this field?

Luigi Ambrosio:
Of course, this depends very much on one's mathematical taste. It would be easy to mention problems from the Clay Foundation, like the regularity for Navier–Stokes.

Alfio Quarteroni:
However, we are interested in your own opinion.

Luigi Ambrosio:
(laughs) Indeed. In my own opinion, I like De Giorgi's perspective. He used to say that difficult problems accompany you throughout your career. It's a good idea to keep trying to solve them. You may not succeed, but it's important not to forget them. At the moment, I would say that the main questions I can think of, within my field, revolve around the regularity theory for least gradient flows. For vector-valued problems in the calculus of variations, there is a well-established notion of convexity called quasi-convexity, which is related to poly-convexity. These ideas appear in mathematical problems in continuum mechanics and are very natural energy functionals from both mathematical and mechanical foundational perspectives. Consider, for instance, functions involving powers of determinants or minors – functions that measure, in some sense, the deformation of a solid in different directions. One significant open problem is that for such functions, there is no evolution theory, gradient flow, or any way to write down a gradient flow. The question remains entirely open, with neither counterexamples nor positive theorems. Many researchers, including De Giorgi and Craig Evans, have attempted to tackle this problem. Evans even suggested it to Alessio Figalli and me a few years ago.

In geometric analysis, a fundamental problem involves the recent theory of Gromov–Hausdorff limits of smooth manifolds. To draw a rough analogy, one might think of the completion of the rational numbers to obtain the real numbers. While a completion in an abstract sense can always be done, the completed object is often difficult to work with unless a concrete realization is provided. For instance, when completing the class of continuous functions with respect to the L^p distance, it becomes challenging to work with the completion unless you know the realization in terms of measure theory and measurable functions.

A complete realization of the closure of Riemannian manifolds is a significant open problem today. This is related to the theory of metric measure spaces. We know that this completion resides within the class of metric measure spaces, but providing a more precise realization or characterization of the closure remains a challenge. Many manifolds with singularities can arise as limits of smooth manifolds, such as convex sets with corners that can be smoothed. However, characterizing these objects is a far-reaching and highly challenging question.

Alfio Quarteroni:
So when you think about realization, you're actually talking about characterization.

Luigi Ambrosio:
Or representation; I would say representation is the right word.

Alfio Quarteroni:
Is there a need in those cases to also have some constructive approaches that allow you to identify classes of elements that belong to that specific completion? Is the constructive aspect relevant in this context, or not necessarily?

Luigi Ambrosio:
At the moment, I wouldn't say that having elementary pieces is necessary.

Alfio Quarteroni:
So, it's about trying to identify the process by which you get to the completion and being able to obtain it.

Luigi Ambrosio:
In some sense, there is little doubt about what the topology should be.

Alfio Quarteroni:
Okay.

Luigi Ambrosio:
Gromov had a beautiful abstract idea of how to define a distance within the class of metric spaces, which might seem like a contradiction – similar to Russell's paradox – because it would make the class of metric spaces a metric space itself. Something should be wrong.

Alfio Quarteroni:
That is indeed a paradox.

Luigi Ambrosio:
Of course, I am oversimplifying, but the idea was to determine how to define a distance in the class of metric spaces. Using this concept or refinements of it, you can talk about Cauchy sequences. Once you have Cauchy sequences, you have the completion.

Alfio Quarteroni:
Sure, sure.

Luigi Ambrosio:
But the realization or representation is another story.

Alfio Quarteroni:

Luigi, I'd like to ask you about the impact of artificial intelligence and machine learning on mathematics. Traditionally, mathematics has been divided into subfields like geometry, algebra, topology, and analysis. However, machine learning has brought about significant changes in nearly every branch of human knowledge. This might seem like an alien concept for mathematicians. How do you perceive the potential interactions, possible collisions, or opportunities that arise from the introduction of such powerful tools originating outside the realm of mathematics?

Luigi Ambrosio:

First, let me clarify that I'm not an expert in this field, though I have attended some talks on the subject. My impression is that the potential of machine learning in fundamental terms is somewhat analogous to that of optimal transport. I see that many people, including those in numerical analysis, are adopting new paradigms related to machine learning. If I understand correctly, there is a new concept of what constitutes an elementary function. I am personally involved in a collaboration with a machine learning expert, who approached me with a problem in geometric measure theory that is intriguing even when considered independently of its applications.

I genuinely see tremendous potential in machine learning to unify and bring people together, who under normal circumstances might not interact. I want to emphasize that I am talking about the fundamental aspects, not just the practical applications. From the perspective of a theorist, I think machine learning offers exciting possibilities.

Alfio Quarteroni:

Your viewpoint is quite interesting. If I understand correctly, you're emphasizing the importance of finding the mathematical foundations behind the collection of algorithms often used with varying degrees of success. What is currently lacking are the rules of the game – understanding why certain methods are successful and why others are not. Efforts to identify the mathematical roots and foundations of this vast field, which encompasses numerous scientific disciplines, should indeed be welcomed. The better we understand these foundations, the more we can generalize, regulate, and control the process. This is an intriguing perspective.

As we are nearing the end of our discussion, I would like to ask you a final question. As the director of the Scuola Normale, you are constantly in contact with young students. Do you have any specific recommendations or advice for young individuals passionate about mathematics?

Luigi Ambrosio:
Okay, it's very difficult. I think this is a difficult question because, as a teacher, I see that the stories of students can be very different. Of course, in some cases, you identify, as I said, some students who are very brilliant and immediately want to do research. But sometimes you find other students who are maybe equally successful, but they prefer to start on fundamental problems. They may appear slower or less brilliant, but that might not be true. So, it's very difficult to give a general suggestion of what the optimal strategy is, so to speak. I would say that the drive for passion is still the main engine. I always like to say that we mathematicians have a kind of innate feeling of what makes a nice problem – a kind of aesthetic feeling, which is difficult to explain to non-mathematicians. So, I say to some friends, the same way you can say, "Well, this is a very nice picture," I am able to say, "This is a very nice mathematical problem." I can't explain to you why. So my suggestion is to be driven by passion and interest. At least that's the way I did it. I started really in a very traditional field and then, out of cu-

riosity and passion, slowly moved into other fields, always keeping in some sense what I learned from the others.

Alfio Quarteroni:

Thank you very much, Luigi. Thank you also on behalf of Springer Nature for giving us the opportunity to talk about a topic – mathematics – that is deep in our hearts.

A. Quarteroni

L. Ambrosio

L. Ambrosio and A. Quarteroni (2022, June 9), Talking about Optimal Transport - Luigi Ambrosio (SNS) interviewed by Alfio Quarteroni (PoliMI and EPFL)

Reference

L. Ambrosio, E. Bruè, D. Semola, Lectures on Optimal Transport, UNITEXT, Springer, 130, 2021

Optimal Transport, Fields Medals and beyond

Alessio Figalli and Luigi Ambrosio

1 Introduction

Welcome to the Springer Math Podcast. This month, we're delighted to host Alessio Figalli, the Director of the Institute for Mathematical Research at ETH Zurich, Switzerland. A distinguished academic, Professor Figalli completed his PhD at the Scuola Normale Superiore of Pisa, Italy, and at the Ecole Normale Superieure of Lyon, France. His research has taken him across France, the United States, and Switzerland. His notable contributions to the theory of

A. Figalli (✉)
Department of Mathematics, ETH Zurich, Zurich, Switzerland
e-mail: alessio.figalli@math.ethz.ch

L. Ambrosio
Scuola Normale Superiore, Pisa, Italy
e-mail: luigi.ambrosio@sns.it

L. Ambrosio, A. Quarteroni (eds.), *Conversations on Optimal Transport*,
https://doi.org/10.1007/978-3-031-51685-6_2

21

optimal transport have earned him numerous accolades, including the prestigious Fields Medal in 2018, and the European Mathematical Society Prize in 2012.

In this enlightening conversation with Luigi Ambrosio, a professor at the Scuola Normale Superiore, Figalli reflects on his initial foray into mathematics, his distinctive career trajectory, and his views on the variance in university systems across countries. He also delves into his time management strategies, the growing significance of optimal transport in recent years, and his approach to problem-solving in mathematics. Originally aired by the Springer Nature webinar series, this interview has been specifically adapted for the podcast format.

Luigi Ambrosio:

It's a pleasure to have this opportunity to converse with Alessio, a cherished former student of the Scuola Normale. Considering that he left the school 15 years ago, I'd like to blend personal inquiries with academic and professional ones. So, let's begin, Alessio. Your career journey is quite remarkable, especially from the outside. Could you share with our audience how you discovered your passion for mathematics and how you came to be affiliated with the Scuola Normale?

Alessio Figalli:

Firstly, I must express my delight in being here, especially being interviewed by Luigi. It's truly a pleasure to revisit and discuss my journey. As to your question, my interest in mathematics wasn't really pronounced until later in my academic journey. I attended a classical high school in Italy, where the curriculum primarily revolved around Latin, Greek, and philosophy. Mathematics was a minor part of what I studied. I enjoyed it, but the limited exposure made it difficult for me to truly appreciate its beauty.

Everything changed when I participated in the Math Olympiads during high school. Unfortunately, math is often taught without context or purpose in school, and many exercises are repetitive and mechanical. While important, this approach doesn't reveal the fun or creative aspects of mathematics. The Math Olympiads, however, exposed me to problems that required non-trivial thinking. At first, they seemed impossible, but as I attempted to solve them and observed others doing the same, I began to understand how to approach mathematical problems.

When I attended the national phase of the Olympiads, I met students from all over Italy who were planning to study mathematics in university and aspired to join the Scuola Normale. Both ideas were new to me. Coming from Rome, a city with its own universities, I had never contemplated studying in Pisa. Yet, the enthusiasm of my peers and the encouragement from university professors made me consider it. They assured me that mathematics was not just about learning old theories but a continuously evolving field.

When I finished high school and applied to the Scuola Normale, I was still unsure if mathematics was the right choice. I remember being very stressed during the examination, particularly the physics part. But I made it, and I decided to give it a try. I'm not certain, because I was young, but I think you were in my oral examination!

Being exposed to university-level mathematics was a revelation. It was abstract, yet it all started to fit into a larger, more exciting picture. However, I must confess, I felt I was lagging behind my peers. Because of the strict grade requirements at Scuola Normale, I was constantly stressed, thinking I might be the one to get kicked out. That fear drove me to study harder, marking the onset of my long mathematical journey.

Luigi Ambrosio:

Your story is indeed unique, and I'll certainly verify if I was on the committee for your 2002 oral examination. It's crucial to debunk the notion that mathematics is akin to a cookbook. When teaching, we always emphasize this to our students to stimulate their interest in mathematics. Now, let's shift to a more professional question. Your career is remarkable in that you quickly moved among various institutions spanning Italy, France, the U.S., and Switzerland. Could you compare these systems, highlighting their peculiar aspects, strengths, and weaknesses? Of course, I understand this is a complex question that might require a lengthy answer.

Alessio Figalli:

That's quite a loaded question indeed. As an undergraduate and Ph.D. student, I began my journey in Italy and then moved to France. I should clarify that I do not consider myself a product of brain drain, where I was forced to leave Italy due to lack of opportunities. In my case, an early opportunity in France presented itself, and I took it. This is not to say that Italy didn't offer me options. I was in Nice and then Paris before moving to the U.S., where I spent 7 years before settling in Switzerland.

From my perspective, I've experienced four distinct systems. Italy and France are fairly similar. Both are fully public systems composed of numerous large universities with vast faculties. These systems might lack competitiveness due to working conditions not being as favourable as those in the U.S. or Switzerland. However, they have their own merits and function well. The only concern is that if funding continues to be reduced, the impact will be increasingly felt.

However, when people suggest that the Italian system has issues and that we should emulate the U.S. or other countries, I'd suggest looking closer to home. France's system is quite similar to ours in Italy but some differences that we could emulate; this would require only a few key changes that could make a significant difference.

One notable aspect of the French system is its continuous hiring process. This might seem trivial, but I believe it's crucial. Talented mathematicians and academics know that each year they have a chance to secure a permanent position. This consistency can be a game-changer. The hiring process doesn't come in waves, where there are periods of no hiring followed by periods of mass hiring.

I remember when I moved in 2007, Italy entered a challenging period with virtually no hiring for several years until recently. France does not experience these hiring freezes. This simple difference is, in my opinion, significant. Additionally, funding in France is more stable, with fewer fluctuations.

The U.S. and Switzerland are quite different in this regard. The U.S. system, even for public universities, primarily relies on private funding. They thrive on external collaborations and substantial grants from these collaborations.

When I was in Austin, a public university in the U.S., only 13% of the funding came from the state, while 87% came from private sources. This implies that the university has to generate income from other sources, such as extensive industry collaborations. Also, a part of the grants obtained by individual academics goes to the university. This system incentivizes hiring those who can attract funding. However, this could be a disadvantage for academics in disciplines where it's more challenging to secure funding, leading to lower salaries. This is just a different system with its own pros and cons, but it's competitive enough to attract many people, despite the high tuition fees.

Switzerland, on the other hand, operates on a publicly funded system. However, only about 20% of the population attends university, leaving 80% to start working right after high school. The country only has eight universities to cater to this smaller demand, making it easier for the state to adequately fund them. This system is based on a completely different culture, where the cut to 20% happens before high school. In Italy, everyone attends high school, as they do in France and many other countries.

Each system has its peculiarities, but the main problem in Italy, in my opinion, is the lack of continuity in financing and hiring. There's a need for consistent support to universities and academic institutions. The main issue is that these aspects are too closely tied to the current government, rather than viewing education as a system that transcends politics.

Luigi Ambrosio:
Now, let's shift our focus a bit. I recently attended the opening ceremony in Helsinki, where, as usual, we watched videos about the Fields Medalists. This time, there seemed to be an even greater emphasis on the importance of collaboration among mathematicians. They didn't just portray these medalists as extraordinary figures, but also highlighted the human side of their work and the significance of family relations. This was an excellent way to depict the life of mathematicians.

Alessio, you're known for your efficiency in everything you do. How do you manage your various responsibilities on professional, personal, and family levels? Do you have a golden rule?

Alessio Figalli:

(laughs) – No. Let's start with your comment. I completely agree. It's crucial to remember that even those who receive awards like the Fields Medal are not isolated geniuses who achieve remarkable mathematics on their own. We are part of a community and benefit from being part of that community. Collaboration plays a crucial role. Looking at all the people present in Helsinki and previous examples, it's evident that significant work is either done in collaboration or is a research project built with others. Even if not direct collaborators, there are other people around the world working on similar topics. Having multiple people working on a topic accelerates its progress. Awards are given for specific results, but nearly all these results wouldn't exist without the contribution of other mathematicians to the topic. It's essential to convey this message. The challenge with prizes is that they're usually given to one person, right? However, I believe that every award also reflects the collective effort of the community working on the topic. Everyone contributes in some way to these accolades.

Moving on to juggling various responsibilities, I believe maintaining a good life balance is crucial for conducting research and for everything in life. Mathematics is a highly creative field, requiring hours of dedicated thinking, sometimes for several months or even years. While this can lead to a certain level of obsession, it's essential not to let this obsession consume us. Finding balance in life, whether it's through family, teaching students, or other activities, can enhance our research abilities.

However, the more roles you have, the less time you have for each. For instance, when I was a student at Scuola Normale, I had all the time in the world. Now, my time has become incredibly limited, and I have to optimize.

I don't believe there's a one-size-fits-all golden rule because everyone is different. For example, I'm good at switching between different tasks quickly, while others might need to stay in a certain "mood" for a few hours to be productive. Each person must find their own way to be more efficient. This could mean dedicating the morning solely to personal work and not responding to emails or meeting students, then reserving the afternoon for meetings, interviews, etc. Or it could be something entirely different.

Becoming efficient is necessary, especially when you have many things to do in your daily life. My efficiency has evolved over time, as the nature of tasks demanded from me has changed.

Luigi Ambrosio:
Yeah, exactly. I fondly recall the days when I was a Ph.D. student at Scuola Normale, where I could spend an entire day pondering a problem. It was magical. But, as life changes, we must adapt. Now, since it was announced that we would discuss optimal transport, I'd like to steer our conversation in that direction. You began working on this topic during your time, our time, at Scuola Normale. Even back then, it was a burgeoning field, with applications in statistics and mathematical physics. Over time, it has evolved to include applications in partial differential equations, geometric analysis, and more recently, machine learning. Are you still interested in this topic? What are your thoughts on its future development?

Alessio Figalli:
I began my journey with optimal transport at Scuola Normale. For those who may not know, optimal transport is a topic that has been around for over 200 years. However, it truly took off as a mathematical discipline in the 1980s. In

the last 35 years, the field has seen tremendous growth for many reasons.

Optimal transport starts with a very tangible problem: how to transport resources or objects from one place to another in the most cost-effective way. It's a classic problem in the calculus of variations, a field beloved by us in Pisa, as it involves minimizing functionals.

While the problem itself is very interesting and has applications in economics (Kantorovich won the Nobel Prize in economics for his work on optimal transport), its growth in mathematics really took off when it became apparent, particularly through the work of Yann Brenier, that optimal transport could be connected to broader concepts.

Subsequently, Robert J. McCann played a key role in demonstrating how this problem could be applied in different areas. As mathematicians, once we start thinking about a problem in abstract terms, we can adapt its models for other uses. For instance, in optimal transport, we're not just transporting material, but we can model it as transporting probability measures, which can represent the objects we're moving.

And once we've reached that point, it doesn't matter whether what we're transporting is something tangible or utterly abstract. Suddenly, we can use optimal transport to move sets and prove isoperimetric inequalities. Back to measures, we can apply optimal transport in probability. We can understand how measures move, whether in flat space or on the Earth, and realize that optimal transport is connected to geometry and curvature. The field has developed in ways most people wouldn't have expected.

Very recently, the connection to machine learning became apparent, something that seems obvious in retrospect. The simplest way to explain this to our audience is probably through image recognition. Images are made of pixels, and if I have two images, I can compare them by calculating the cost of transporting the pixels from one image to the other. This allows me to determine, for instance, if two images are similar, perhaps representing the same object. For example, machine learning started booming when it began recognizing digits and postcodes. Optimal transport can be used for that. You can read a bunch of numbers and then try to recognize them.

I'm not suggesting this is the most efficient way, but I'm trying to give a sense of why optimal transport naturally has such broad applications, because you can transport everything. This is something the computer science community has fully recognized. There's increasing interest in optimal transport at computer science conferences. People working on optimal transport are even moving to private companies, which are now seeking expertise in this area.

There's been a massive boom in the field, which is quite surprising. From my perspective, when I began working on optimal transport, I was very excited by the applications it had. In my work, I used optimal transport to investigate isoperimetric inequalities – informally speaking, to understand how the shape of crystals changes when a force is applied to them. This is really about geometric inequalities. I also applied it to semi-geostrophic equations, which are equations used in meteorology. These are extremely complex partial differential equations, but a clever change of variables transforms this complicated problem into another where optimal transport naturally arises.

Without getting into details, a key question that emerged from optimal transport was about the regularity of optimal maps. As a mathematician, you might ask: if I'm transporting one density to another and I know that these distributions are 'nice', can I say that the optimal transport map between them will be 'nice' as well? How is this 'niceness' transferred from the density to the map? This is what we mean by regularity.

This is a subject I've studied extensively, and it played a crucial role, particularly in my receipt of the Fields Medal. Moreover, moving from country to country exposed me to different mathematics. Optimal transport was very active in Italy and France.

When I moved to the U.S., particularly to Austin, which is a hub for free boundary problems, I started to explore an entirely different area. Free boundary problems involve understanding the separation between phases. For instance, if you have ice in water and the ice starts to melt, you want to understand this melting phenomenon. You want to study the boundary between the ice and water and how this surface separating the two phases evolves over time. I began working on this subject, which perhaps took me a bit away from optimal transport.

Recently, however, I've started to regain interest in optimal transport, especially in its more applied questions, which often lead to purely theoretical, highly mathematical problems. I believe there's still much to be done, particularly as new connections to machine learning are bringing up new questions in optimal transport. So yes, I am indeed interested in what's happening in the field right now. I find it very exciting.

Luigi Ambrosio:

Indeed, the history of optimal transport is fascinating, particularly from the perspective of the evolution of ideas. As you mentioned, the problem was first proposed during the French Revolution, then somewhat forgotten, only to be rediscovered by Kantorovich around the 1940s. The subsequent branching into various fields occurred almost independently, with contributions from Yann Brenier in France,

Robert J. McCann in Canada, and Felix Otto in Germany. It's intriguing how at times, a problem seems to be "in the air," and these independent efforts start to converge.

I'd also like to touch on the connection between optimal transport and machine learning. While I'm not an expert, I've attended workshops and lectures on the topic. I get the sense that there's an analogy between the two, not only in the sense that optimal transport can be applied within machine learning and vice versa. Like optimal transport, machine learning seems to have the potential to influence various other fields like PDEs and optimization, thus introducing new paradigms.

Is this just my perception, or do you also see machine learning having this kind of potential—setting aside the validity of its applications—purely at a theoretical level?

Alessio Figalli:
Absolutely, you're right. This is more of a general question, but machine learning, despite existing for a while, has only recently experienced a real boom. There are still many technical, even basic questions about why many things work as they do that remain unanswered. So, there are indeed two similarities. One is the practical connection between optimal transport and machine learning, and the other is the broader, theoretical connection. Yes, I share your sentiment.

Luigi Ambrosio:
Now, let's move on to a more personal question. From the outside, it seems like everything comes easily to you. I've only ever seen you tired once, at the end of your Ph.D. I remember you telling me during the first year of your Ph.D. that you were a bit worn out. There was a similar legend about Ennio De Giorgi, that everything was easy for him. However, during De Giorgi's farewell ceremony, his

sister Rosa contradicted this legend. She recalled times when Ennio worked hard on a problem in isolation. Of course, I can attest to De Giorgi's exceptional speed on many occasions. What can you say about this in your case?

Alessio Figalli:
Well, I wouldn't put myself on the same ground-

Luigi Ambrosio:
What about you?

Alessio Figalli:
I wouldn't say everything comes easily to me. Mathematics is an art, and regardless of how naturally some aspects might come to some people, we all face challenges and difficulties. Frustration is a common part of the process. My ability to switch between different problems with ease has always been a benefit to my research career. For instance, the problem of applying optimal transport to semi-geostrophic equations, which you gave me in 2005, wasn't solved until 2012, with the help of Guido De Philippis. So, it certainly wasn't easy. It took 7 years of trial and error. In fact, all our initial attempts were off the mark, as the final proof we found took a completely different direction.

Luigi Ambrosio:
That was in Oberwolfach, right?

Alessio Figalli:
Yes, it was in Oberwolfach in 2012 that we had the key idea. However, it would be wrong to say that the previous 7 years were wasted. Attempting things and experiencing failure teaches you a lot. Sometimes, it leads you onto the right

path. Without that previous thinking, we wouldn't have made it.

But what I want to emphasize is that while that was a success story, there are many problems that we try to solve and fail. We all have a list of such attempts. We tend to remember the moments when someone immediately had the right idea. That of course impresses us. It is impressive when someone suddenly comes up with an idea. You react with, "Oh, wow, how did you come up with that?". It can lead to people acquiring legendary status. But in reality there are many situations where someone presents a problem to me and I have no clue how to solve it. That's part of research in general, mathematics in our case. Our research is challenging because we are trying to do something that people couldn't do before us, which means that in general it is difficult. I think it is important to stress this fact, because otherwise, many times when we get stuck, we might think, "oh, I'm not good enough to do this job. What should I do?" In reality, we all pass through this, and the number of problems we actually solve is way less than the ones we tried to solve. In the end, that's how you build your career. You try to do your best and you solve problems that you're interested in, which took your energy many times, but it's normal not to be able to solve many of them. So, I wouldn't consider myself a legend.

Luigi Ambrosio:
Yes, our time is almost up. Let's conclude with a local, Italian question. I've been very involved in discussions with the Ministry of Education, emphasizing that many top students and researchers, particularly from Scuola Normale, leave the Italian system. This brain drain is a significant concern for many of us.

Specifically, for Scuola Normale, we make a considerable investment in our students. We do see a return, especially in the form of an international network that we're building. However, we should certainly aim to improve this situation.

We've already discussed the reasons for this diaspora, so perhaps we can focus more on recommendations. What would you suggest to improve the situation?

Alessio Figalli:
Indeed, Scuola Normale is an example close to our hearts, given your present role as the Rector and my history as a former student. It's certainly regrettable that many students leave, but this is part of a larger issue. Every year, thousands of brilliant Italian students move away.

Focusing on the academic world, I believe it's healthy for students to leave at the early stages of their careers. The Italian system could benefit greatly from this if it could bring them back. This movement, particularly at the Ph.D. and postdoctoral levels, would be excellent if it led to people returning to Italy. The issue is that while people in the past were hesitant to move, it has now become very easy for us. Once you experience a potentially more straightforward academic system, perhaps one that offers positions at early career stages and has a well-structured path without surprises, it can be hard to leave that system.

Luigi Ambrosio:
The problem of securing a position when you're young is a significant concern. Opening enough junior positions to allow the best people to stay and build their careers could be one solution.

Alessio Figalli:

Absolutely. For instance, if I'm inside the system and want to be promoted from associate to full professor, the university needs to have enough funds allocated for this promotion. This often results in internal competition among the 10 associate professors who all aspire to become full professors. This is not a healthy situation. Promotions should be evaluated on individual merit, not turned into a competition between colleagues. This is particularly true because the idea is for you to grow within your university, not to move again.

So, it all boils down to having sufficient resources to allow deserving individuals to be promoted. This is one of the major factors that discourage people. And once you've settled down outside of Italy, perhaps even started a family, it becomes too late to move back. After a certain age, moving becomes challenging. Therefore, we'd want people to return at an early stage. Ideally, people would go abroad in their early 20s but then return by their early 30s. This is a rough timescale, of course.

Luigi Ambrosio:

So, there's a time for mobility and a time for -

Alessio Figalli:

Yes, for stability. Also, those with a solid technical education have many opportunities in the private sector. If you have a Ph.D. in mathematics, you stand a good chance of being hired. I see it here; people move on to work at Google, IBM, Apple – these international companies are attracting a lot of mathematicians. Perhaps this is something that's lacking in Italy on a national level: adequate working conditions for people with advanced training in scientific studies. This might be another factor motivating people to leave, unfortunately.

Luigi Ambrosio:
Thank you very much, Alessio.

Alessio Figalli:
Thank you.

Ambrosio and Figalli, Lyon, 2015

A. Figalli and L. Ambrosio (2022, September 7), Optimal Transport, Fields Medals and beyond - Alessio Figalli (ETHZ) interviewed by Luigi Ambrosio (SNS)

References

L.Ambrosio, E.Bruè, D.Semola: *Lectures on Optimal Transport.* Springer Unitext Series, 130, 2021.

L.Ambrosio: *Calculus, heat flow and curvature-dimension bounds in metric measure spaces.* Proceedings ICM 2018, Vol. I, World Sci. ed., 301-340 (2018).

L.Ambrosio, N.Gigli, G.Savarè: *Gradient flows in metric spaces and in the space of probability measures.* Lectures in Mathematics ETH Zürich. Birkhäuser Verlag, Basel, second edition, 2008.

J.-D.Benamou, Y.Brenier: *A computational fluid mechanics solution to the Monge-Kantorovich mass transfer problem.* Numer. Math., 84 (2000), 375-393.

G. De Philippis, A. Figalli: *Partial regularity for optimal transport maps.* Publ. Math. Inst. Hautes Etudes Sci., 121 (2015), 81-112.

A. Figalli: *Regularity of interfaces in phase transitions via obstacle problems.* Fields Medal lecture. Proceedings of the International Congress of Mathematicians—Rio de Janeiro 2018. Vol. I. Plenary lectures, 225-247. World Scientific Publishing Co. Pte. Ltd., Hackensack, NJ, 2018

A.Figalli, J.Serra: *On the fine structure of the free boundary for the classical obstacle problem.* Invent. Math., 215 (2019), 311-366.

A.Figalli, F.Glaudo: *An invitation to optimal transport, Wasserstein distances, and gradient flows.* EMS textbook, EMS Press, 2021.

R.J. McCann: *A convexity principle for interacting gases.* Adv. Math., 128 (1997), 153-179.

F.Otto: *The geometry of dissipative evolution equations: the porous medium equation.* Comm. Partial Differential Equations, 26, 101-174, 2001.

F.Santambrogio: Optimal Transport for Applied Mathematicians. Calculus of Variations, PDEs, and Modeling. Progress in Nonlinear Differential Equations, 87, 2015.

A. M. Vershik: *Long History of the Monge-Kantorovich Transportation Problem.* The Mathematical Intelligencer, 2013.

C.Villani: *Optimal transport. Old and new.* Grundlehren der Mathematischen Wissenschaften, 338. Springer-Verlag, Berlin, 2009.

From moving masses to bending spaces: an excursion in metric geometry

Nicola Gigli and Camillo De Lellis

1   Introduction

Welcome to the Springer Math Podcast. In this month's podcast, Camillo De Lellis, a researcher at the Institute for Advanced Study in Princeton, converses with Nicola Gigli from the Scuola Internazionale Superiore di Studi Avanzati in Trieste, Italy. They delve into Nicola Gigli's personal journey in and out of mathematics, discussing his path to the topics of his research and his enthusiasm for them. In

N. Gigli (✉)
SISSA, Trieste, Italy
e-mail: ngigli@sissa.it

C. De Lellis
School of Mathematics, Institute for Advanced Study, Princeton, NJ, USA
e-mail: camillo.delellis@ias.edu

L. Ambrosio, A. Quarteroni (eds.), *Conversations on Optimal Transport*,
https://doi.org/10.1007/978-3-031-51685-6_3

this conversation, they also explore the connection between the concepts of optimal transport and the curvature of space, a discovery that has given rise to a flourishing research field at the intersection of multiple areas of mathematics. Originally aired by the Springer Nature webinar series, this interview has been specifically adapted for the podcast format.

Camillo De Lellis:
Thank you very much, welcome everybody. I am Camillo, the interviewer. It's a great pleasure to be here to interview a brilliant mathematician like Nicola, who's also a long-time friend. So, Nicola, let me start with a very classical question. When did you discover that mathematics was going to be one of your lifetime passions? I mean, is there one particular decisive episode when you thought, okay, I want to become a mathematician?

Nicola Gigli:
Well, first of all, Camillo, let me say, it's a special honour being interviewed by you. We've known each other since the early stages of our careers. I think we will touch on this sometime later during this interview. As far as I remember, I've always loved playing with numbers and adding things up. My mother recalls that when I was 2 or 3 years old, I was already counting and adding up, two plus two, two plus three, and so on. So, in a sense, it's something I always sort of had. My family was good at stimulating me. My grandfather, for instance, would let me sit on his knee and play with numbers. He taught me the casting out nines trick, which you use to check arithmetic computations quickly. For me, these sorts of things were like magic. Of course, it took 10 years or more to actually understand how that could work. But I received stimulus at a very early stage

from my family, despite the fact that they are not university-educated. So yes, when I decided to pursue mathematics, it just came naturally. I had this great passion, and it was a natural thing to try.

Camillo De Lellis:
Wonderful. Nicola, you know there was a second motive to ask this. Kolmogorov once famously said that the emotional maturity of a mathematician coincides with the moment at which the mathematician showed an inclination for mathematics. So, you answered that. Two or three years, okay?

Nicola Gigli:
I didn't know this.

Camillo De Lellis:
You didn't know this?

Nicola Gigli:
Is it true?

Camillo De Lellis:
Yes, it's true. I read it somewhere. It was actually done on purpose to tease another famous Soviet mathematician who showed an inclination for mathematics at the age of three or four.

Nicola Gigli:
[laughs] I see, Okay. I don't know what to say.

Camillo De Lellis:

Well, okay. So we are two emotionally three-year-old people then, I guess.

Nicola Gigli:

Yeah, in fact, I guess I suspected.

Camillo De Lellis:

Okay. So how influential were...

Nicola Gigli:

I'm not sure – is this good or bad? I'm confused. Well, let's leave it at this.

Camillo De Lellis:

Well, I don't know. I think it says, somehow, that kids are very fresh. So, you can make an argument for that. How influential were the first years at university on your career and what are your best memories of that time?

Nicola Gigli:

Okay, perhaps speaking of my career, if you allow me, I want to mention what happened before coming to university. I guess we both grew up in what is basically the prehistory of mankind, before the internet, right? At that time, it was not that easy to find sources and stimuli for something like math, especially coming from a small city like we did. I want to mention that before arriving at university there was this very impactful moment of the Mathematical Olympiad and the whole community of young boys and girls who loved solving riddles basically, but special kinds of riddles. That was extremely impactful. That was a moment where finally you met somebody who has your own taste and interests.

Camillo De Lellis:
And if I may interject, I think the Italian Mathematical Union does a wonderful job to actually make it very widespread, right?

Nicola Gigli:
Yes, I agree.

Camillo De Lellis:
It's always mentioned that some country or another did fantastically well. But what I found out when I actually went outside is that while many countries actually get the top ranking and get the medals and so on, not that many have the widespread movement that we have in Italy. We involve essentially all high schools, or at least as many high schools as possible.

Nicola Gigli:
And they did a lot over the years. They are doing a better job than when we were young.

Camillo De Lellis:
The involvement is just absolutely outstanding and I think it's one of the great efforts that the Italian Mathematical Union does. It's not easy to fund that activity. But okay, sorry for interrupting you.

Nicola Gigli:

No, sure, I mean, it's good to praise a good activity, so absolutely, I agree. Where were we? So after that there was Pisa, and I studied at the Scuola Normale, and that of course put me in a very special environment where I was surrounded by extremely talented guys of the same age who I could have a beer with and discuss math. Typically, the two things can be done together. So they were very, very impactful. I remember when I started, in the very first year of undergraduate studies, I was leaning towards studying algebra or algebraic geometry, so something completely different from what I'm actually doing research on now. And that's mainly because there was this amazing professor, Enrico Arbarello, in Normale doing his classes, including a course that he gave when I was in the first year, which was basically a course on Galois theory. You know, analysis is interesting, but something that I had already heard about in high school, so, I asked, what more can there be? In my mind at the time, I thought once I can differentiate and integrate, that's all analysis can give, but then I followed these lectures on Galois theory. This duality in math is, I think, a bit surprising, and I really loved it. I was going in that direction, and then, well, I guess, you, Sunra Mosconi, Paolo Tilli, and other friends that I used to hang out with at university pushed me toward analysis. So, I did my first serious study with Paolo Tilli at the colloquium. We had to do this special sort of bachelor's thesis in our third year of university. He suggested that I go with Luigi Ambrosio for my undergraduate study. I would say I was first in love with algebra, but then after chatting with friends about analysis, I ended up focusing on that.

Camillo De Lellis:

So, coming back to those days, at a certain point, you weren't sure of becoming a professional mathematician, right? You actually decided to try and find a job in the private sector. Then you decided to come back. What do you think of that experience and how has it changed you?

Nicola Gigli:

Yeah, thanks for this question. As I always say, I don't regret having gone, and I certainly don't regret having come back to academia. Basically, it was a time where I was concluding my PhD. I felt like it was time for big decisions. I also did my PhD in Pisa. So even from a personal standpoint, after living 7 years or so in one place, it was time to move. I really was not sure anymore about what to do because, up to that time, I had basically only done math. I felt I had other talents and something else that I could have been able to do. I wanted to try something completely different, so I went to the private sector and I started working. In fact, even here, the influence of friendship had a role – our common friend, Francois Constantino, also had a similar career path and gave me some suggestions. I was not the first to do something like this. I tried working in a management consulting company, McKinsey, which is a completely different work from that of a mathematician. When I think about it, it's really kind of complementary because when one works in that sector, one has to deliver products in some sense, every day, sometimes even more than once a day, whereas our job as a researcher typically has a span of months or years before one tries to attack a serious problem, something that we start, but who knows when we will finish. It was very nice. I loved the job. Although at some point, even there, due to the nature of the job, I was supposed to make a decision: if I was going to continue there, I should have been in the business for another 3 or 4 years. But I was not ready to make that sort of commitment, so, I decided at some point that maybe I could do some trading in investment banks, investment funds and whatever. So, I started sending out my CV in that direction. Then 1 week later, Lehman Brothers failed, it was 2008, and I took it as a sign of, you know what, okay, let's just go back to doing math and studying. But for those younger people that are listening, or who will listen to this, I certainly advise you to give it a try if you have in mind the idea of seeing what's outside, because there are a lot of interesting jobs. For me, I really felt a change of pace. When I was doing a PhD and certainly even more when I was an undergrad, math was more of a game in some sense. Maybe it was the influence of mathematical olympiads or whatever, but it was just for fun. When I came back after McKinsey, I really felt I was a professional mathematician. Even the first day back to the office, or at conferences, despite the fact that I was away from the community for 3 or 4 years, I don't remember exactly how long, it depends on how you count, I was really a way better mathematician than I had been before. So maybe this would have happened anyway, with getting older and whatever, but for me, I felt a change of pace.

Camillo De Lellis:
I think it would have happened anyway, but okay, you know that.

Nicola Gigli:
It could have.

Camillo De Lellis:
It's actually also good advice for all the young people who are at the beginning of their undergraduate studies, right? I mean, because there's this idea that you're getting a degree in mathematics, what can you do? High school teachers, most people, actually have this idea, which is very outdated. A degree in mathematics opens the door for a lot of jobs, not only in teaching, not only in academia, but also in the private sector, in industry, in lots of ways, and I think this is not so well known.

Nicola Gigli:
Yes, at least in Italy.

Camillo De Lellis:
I can certainly say that, at least in Italy, when we were growing up, this was the case. Now, let's delve a bit more into mathematics. How did you come to be interested in the research topics you are working on today? Perhaps you've already touched on this a bit. Your taste for mathematics evolved as you aged. How much did it change? You've already mentioned that it changed quite a bit at the beginning, but I'm mostly interested in how you decided on your current focus. How did you decide, "This is what I really want to do"?

Nicola Gigli:
That's a good question. I remember when I attended one of the first conferences around 2008 or 2009 after returning to academia. Cédric Villani was there, giving a series of lectures on his recent results with John Lott and the parallel ones of Sturm concerning lower bounds on the Ricci curvature on non-smooth structures. The possibility of

discussing geometry in this setting amazed me. I was impressed by how much information one could extract from a seemingly simple set of axioms. However, it was clear at the time that there was more to be said and there was a lack of tools to study these spaces. I became fascinated with the idea of building mathematical tools to study these objects. I was in a fortunate position. Sometimes luck plays a role in life. I had experience studying gradient flows in the metric setting, thanks to a book that with Luigi Ambrosio and Giuseppe Savarè we wrote while I was doing my PhD. It turned out that this concept played a crucial role in the study of these spaces. By sheer luck, I had a set of tools at my disposal that I could use to study these spaces. I believe it's a combination of something that inspires you and the ability to say something that others might find interesting. I'm not sure of the exact ratio, whether it's 50-50 or 70-30, but it's definitely a mixture of these two elements.

Camillo De Lellis:
Would it be fair to say that over the years, you've transitioned from a problem-solving approach, like during your math Olympiad time, to a theory-building approach?

Nicola Gigli:
Absolutely. I love theory building. If I had to describe the kind of math I do, it's more theory building than problem solving. Of course, one has to be cautious with this because theory building for its own sake can be useless. You might create a beautiful theory that no one cares about.

Camillo De Lellis:
And I think you solve problems while you're building theory too, right?

Nicola Gigli:

Yes, that's true. But if I had to choose, I would lean more towards theory. However, in doing so, you have to demonstrate that the theory has value. So, you have to come up with a good theorem.

Camillo De Lellis:

Since we're on this topic, can you describe in simple terms the link between optimal transport on one side and the curvature of spaces on the other, which can be applied even to situations where the space is too singular to understand curvature in the way it was understood in the nineteenth century?

Nicola Gigli:

Sure, with pleasure. This goes back to the title of what I want to talk about today. There's a very famous analogy in metric geometry. If you're dealing with convex functions, say a convex function defined on R or a space in multiple dimensions with values in R, you can define convexity in two ways. Either by asking the Hessian to be non-negative or by asking for the convexity inequality to be in place. While these are equivalent for smooth functions, the second one, the convexity inequality, can be formulated without requiring any a priori smoothness on the function itself. So, in some sense, it captures the true essence of convexity, if by true essence we mean, *a la* Bourbaki, something that doesn't require additional data like smoothness. Now, it's an amazing fact that this principle also applies to curvature. Once you realize that, for Riemannian manifolds, curvature influences how fast geodesics spread out or converge. When the curvature is positive, geodesics tend to approach one another. When the curvature is negative, geodesics tend to move farther apart. In fact, one can speak about sectional curvature bounds, here there is no optimal transport, by studying the convexity of the distance function from a given point. You have a Riemannian manifold, you fix a point, you look at the distance function from that point along other geodesics, and depending on whether this function is convex or concave, you can understand lower and upper sectional curvature bounds. Now, the beautiful idea of Lott and Villani on one side, and Sturm on the other, comes into play. Basically, Ricci curvature is the average of sectional curvature. So, Ricci curvature doesn't look at the convexity of the distance along a single geodesic, but considers many geodesics together. And when you want to average mass, probability measures come into play. When you want to move probability measures along geodesics, optimal transport comes into play. So optimal transport gives you a way of averaging curvature along all possible directions and therefore a way to get information about Ricci curvature rather than about sectional curvature. This is a simplification, of course, but the point of view is this: you can express lower Ricci bounds in terms of the convexity of certain functionals over the space of probability measures equipped with the optimal transport geometry. Every time I have a space where I have geodesics, and I have a function on that space, I can always wonder whether the function is convex or not. I don't need differentials, I don't need a Hessian, I just need the convexity. That is the key point that allows the non-smooth theory to kick in.

Camillo De Lellis:
Just for the record, what you described at the beginning, for sectional curvature, is the idea of Alexandrov geometry.

Nicola Gigli:
Exactly, Alexandrov geometry, and the theorem that I mentioned about convexity is the Toponogov theorem.

Camillo De Lellis:
Which is somehow the foundational point of view for Alexandrov geometry.

Nicola Gigli:
Absolutely. And then of course, we should mention Gromov. We're speaking about geometry...

Camillo De Lellis:
Yes, of course.

Nicola Gigli:
Gromov should always be there. He's the person who really proposed studying these non-smooth structures as a means to gain new information about smooth Riemannian manifolds.

Camillo De Lellis:
But there's a whole school on that, of course. So, what do you like most about these topics and what do you like least about them?

Nicola Gigli:
What I love the most is creating things. Whenever you give a definition, you're creating something that wasn't there before. Doing this is something that I like. Creating tools that allow someone, even yourself, to better study a geometry that was already given before, that's what I love the most. What I like the least is, I guess, there's a certain amount of work that we as mathematicians should do, especially when we introduce a new concept: there's a bit of work involved in checking its compatibility with what was previously available. You want to make sure that what you're doing is

compatible with this, this, this, and that, or if it's not compatible, why it's not, and so on. That's a bit less exciting for me. But still, this power of creation is what I love the most.

Camillo De Lellis:
Now, the most important question for me, as the interviewer. You have quite a few younger collaborators and people that you're mentoring, like postdocs, PhD students. How do you choose the problems that you give them?

Nicola Gigli:
That's a good question.

Camillo De Lellis:
I'm going to take notes.

Nicola Gigli:
It's a difficult question. It's particularly important here in SISSA, because we're a PhD school. Giving PhD theses is essentially what we do, so that's crucial.

Camillo De Lellis:
Very successfully though.

Nicola Gigli:
It's easier when the field in which you're doing research is young because in this case, many questions are not just big, unapproachable problems. What I typically do is think of a problem where I sense there could be something, without really being sure that there is actually something, or even if there is something, whether it's the way that I thought it was. Then I start discussing it with the student to whom I

proposed the topic. If things work out as I originally thought, then we have a result. If they don't, we typically learn something in the process. But you're right, it's a very delicate question because you can't give problems that you already know how to solve because that's not a research problem, that's an exercise. And you also can't give extremely hard problems because then there's the risk that after 3 or 4 years, the duration of the PhD at your institution, the student has nothing.

Camillo De Lellis:
Do you keep a list?

Nicola Gigli:
No. Maybe I should actually.

Camillo De Lellis:
Now I gave you an idea.

Nicola Gigli:
If I kept a list, then I would want to study this and that, and I would be unable to keep myself from thinking about the problems, but maybe I should.

Camillo De Lellis:
What is the theorem that you're most proud of? Your theorem.

Nicola Gigli:
What if it's not a theorem, but a definition? Can it be a definition?

Camillo De Lellis:

Well, you're cheating, but okay, fine. If it's a definition, it's fine. It's a theorem-definition, maybe.

Nicola Gigli:

Sorry?

Camillo De Lellis:

It's a theorem-definition. Who was the one who introduced this? Grothendieck? Or am I wrong? A definition-proposition?

Nicola Gigli:

Say it again, sorry?

Camillo De Lellis:

I mean, what did he call it? A definition-proposition. I mean, one of those things.

Nicola Gigli:

Yes, I understand. What I'm really most proud of is really a definition, not a theorem-definition, and it's the definition of an infinitesimally Hilbertian space. So, to make a long story short, in the class of spaces introduced by Lott, Sturm and Villani, not only Riemannian manifolds were included, but also Finsler manifolds. This is good from some perspectives and less good from others. It was a natural question to try to understand whether one can single out more Riemannian-like spaces among these non-smooth structures, and of course, given that you don't have tangent spaces that is tricky to do. The definition that I gave concerns properties of Sobolev functions on those spaces. That is really what I'm most proud of because a few years of work was necessary in order to give a meaningful definition and another few years of work were needed to convince the community, with geometric results, that the definition was actually meaningful and not just something written in a paper.

Camillo De Lellis:
What do you see in the future for the topics that you're studying?

Nicola Gigli:
Good question. I think, what I would love, and this, at least in some subfields, is already happening, is that the community will become confident in the possibility of using certain tools from non-smooth geometry, off the shelf, to tackle possible problems even in, say, the smooth Riemannian or Lorentzian setting (more recently I have been getting interested in Lorentzian geometry). So in other words, you can be a guy that wants to work just with smooth functions but still you surely know that there are Sobolev spaces somehow, somewhere, and that sometimes they can be useful in order to get existence results for some PDEs or to get a certain estimate or whatever. In some sense I'd love it if a similar sort of thing could happen with these curvature bounds, either an Alexandrov, a Ricci or a scalar bound. It's a big question to understand the non-smooth counterpart of scalar curvature bounds. The whole business of Lorentzian geometry is extremely green as of now. Only basic definitions are given, so there's still a lot to do. Perhaps my answer was more on the culture of math than on the theorems that I expect, but I hope I gave a satisfactory answer.

Camillo De Lellis:
Wonderful. So let me just ask you a final question. Since I am a nasty guy, there's a type of question that I always hate when I am asked. It is the type of question like "what would

you advise a young person to do?" in this or that situation. But, you are the one answering, so I can ask you a nasty question, right? What would you advise a young mathematician to do? I mean, a young wannabe mathematician who hasn't yet decided what topics to tackle. How would you advise this person to choose a direction, say, in her or his mathematical endeavor?

Nicola Gigli:
That's a good question. First of all, I would suggest this person to be open-minded and not think that when you are an undergrad you are deciding the topic of research that you're doing for your whole life. I mean, don't think that if you start doing analysis in the third or fourth year of university, then you will do that forever and ever.

Camillo De Lellis:
That's great advice.

Nicola Gigli:
So, make decisions, they are important, but it's not...

Camillo De Lellis:
You're still very young.

Nicola Gigli:
Yeah, you're still very... I mean, you basically...

Camillo De Lellis:
You're a baby.

Nicola Gigli:

You should acknowledge at some point that you really don't know enough to make a sound decision. Go in the direction that you like the most, and don't be afraid of keeping the door open to other directions. In this sort of discussion, I think it is extremely important, especially if you're young, but even if you're not so young anymore, to speak with people, both your peers and your seniors. Math is really something that is done in a community in some sense. Nobody's an island – communicating with people, trying to understand what they do, what they like, what they're interested in, that's something extremely important and this is good advice even to more senior colleagues. Try to speak not just with the person in your niche, otherwise the setting can become asphytic, at least that's my taste. Speak with a variety of people if possible.

Camillo De Lellis:

Yeah, you did a great job actually, so. I will steal this answer for whoever actually asks me the same. Okay, wonderful.

Camillo De Lellis teaching in a classroom

Nicola Gigli

C. De Lellis and N. Gigli (2023, July 10), From moving masses to bending spaces: an excursion in metric geometry - Nicola Gigli (SISSA) interviewed by Camillo De Lellis (IAS)

Correction to: Talking about Optimal Transport

Luigi Ambrosio and Alfio Quarteroni

Correction to:
Chapter 1 in: L. Ambrosio, A. Quarteroni (eds.), Conversations on Optimal Transport, https://doi.org/10.1007/978-3-031-51685-6_1

The original version of the chapter "Talking about Optimal Transport" was inadvertently published with incorrect captions for the images on page 18. It has been corrected now to read as follows and the same has been updated in the chapter.

The updated versions of this chapter can be found at
https://doi.org/10.1007/978-3-031-51685-6_1

L. Ambrosio, A. Quarteroni (eds.), *Conversations on Optimal Transport*,
https://doi.org/10.1007/978-3-031-51685-6_4

A. Quarteroni

L. Ambrosio